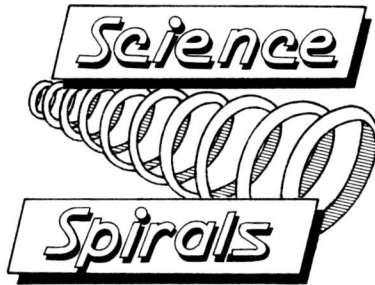

*Science Spirals*

# Magnets

Julie Fitzpatrick
Illustrated by Sara Silcock

Silver Burdett Company
Morristown, New Jersey

**Library of Congress Cataloging in Publication Data**

Fitzpatrick, Julie.
  Magnets.

   (Science spirals)
   Includes index.
   Summary: Gives instructions for simple experiments
that answer questions about the characteristics of
magnets and how they work.
   1. Magnets — Experiments — Juvenile literature.
[1. Magnets — Experiments. 2. Experiments] I. Silcock,
Sara, ill. II. Title. III. Series: Fitzpatrick, Julie.
Science spirals.
QC757.5.F58   1985     538'.2'078     84-40838
ISBN 0-382-09061-6

First published in Great Britain in 1984 by
Hamish Hamilton Children's Books
Garden House, 57-59 Long Acre, London WC2E 9J2

Designed by Linda Rogers Associates

Adapted and published in the United States, 1985, by
Silver Burdett Company, Morristown, N.J.

ISBN 0-382-09061-6
Library of Congress Catalog Card No. 84-40838

# Introduction

You probably have some magnets holding paper to your refrigerator. Have you ever wondered how they work? Have you looked at a compass? Have you asked anyone how compasses work?

This book will help answer some of these questions. It will also show you how to do some easy experiments with magnets. And it will show you how to have some fun with them, too.

# Remember

Keep magnets away from watches and calculators.
The power of the magnet may stop them from working properly.

# Equipment you need for experiments in this book

Bar magnets
Horseshoe magnets
Things made of metal:
   paper clips, nail file, bobby
   pins, pins, nails, screws,
   keys, and coins
Paper
Pencil
Pencil sharpener
Ruler
Eraser
Scissors
Clay
A large plastic bowl

Tape
A small box
A piece of cloth
A plastic bottle
A clear plastic container
Some small stones or shells
String
Thread
A wooden block
A packet of needles
A ceiling tile
A plastic tub
A compass

# What things will a magnet attract?

There are all sorts of magnets.
They are different shapes and
sizes but they all have
a special power.
What can magnets do?

They can pick up some things.
We say that the magnet
attracts these things.
Try a magnet around the room.
What can you find that is attracted
to the magnet?

Try a magnet on some large things.
Will it stick to them?
Find out which parts of
a bicycle are attracted
to the magnet.

Collect as many different things
as you can.
Make a guess which ones
will be attracted to the magnet.
Sort your things into these two sets.
Which things do you think
will be attracted?

will be attracted

will not be attracted

Now test each one with a magnet and
see if you were right.

Did you have any surprises?

| were attracted | were not attracted |
| --- | --- |

Look at the things which were
attracted to the magnet.
What are they all made of?

They are made of metal, but
there are other kinds of metal.
Do you think that a magnet
will attract other kinds of
metal as well?
A clue for you:
look back at the things
which were not attracted.

A magnet will not attract
all kinds of metal.
It only attracts metals
which have some iron in them.
These are called magnetic metals.
Metals which are not attracted
are called non-magnetic metals.

Magnetic metals

Non-magnetic metals

# The Flow Chart Game

This is another way of sorting the things you have collected. Put each thing on 'Start' and follow the Flow Chart.

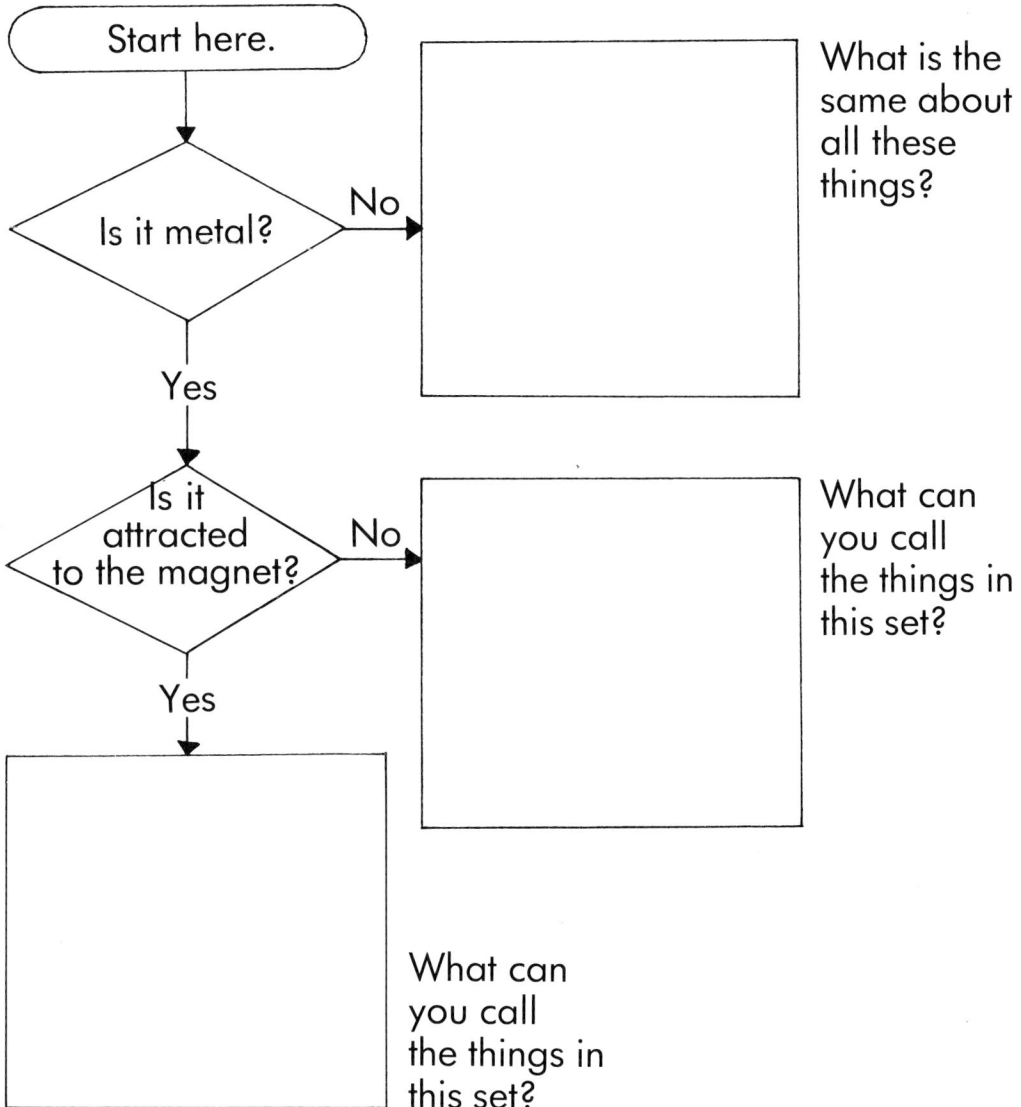

Start here.

Is it metal? — No →

What is the same about all these things?

Yes

Is it attracted to the magnet? — No →

What can you call the things in this set?

Yes

What can you call the things in this set?

## Which part of the magnet is best for attracting things?

Which part of the magnet
have you been using
to attract things?
What happens if you use the
middle part of the magnet?
Use a paper clip to touch all over
your magnet.
Can you feel the paper clip
being attracted more at the ends
or in the middle of the magnet?

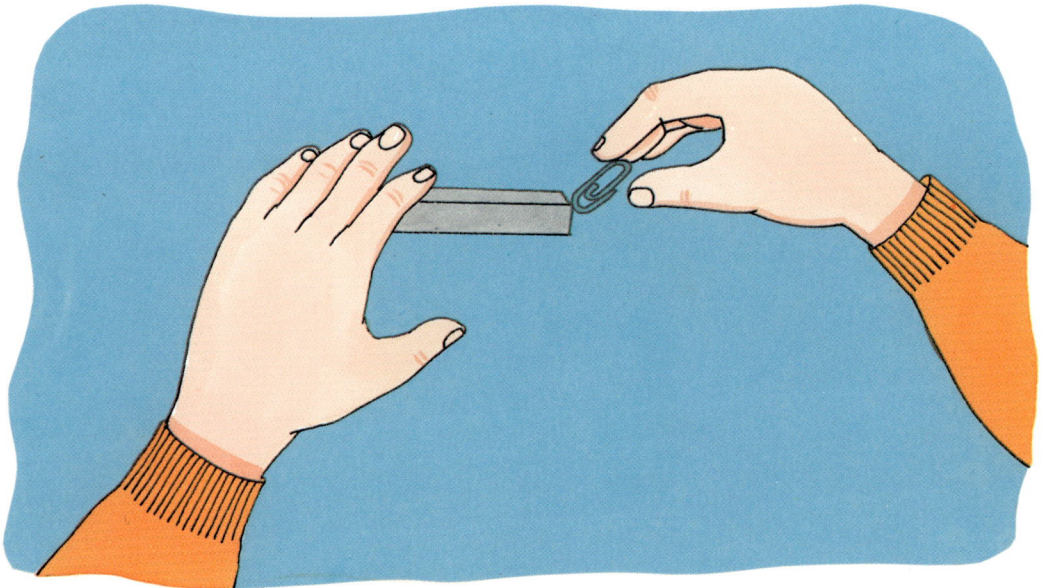

The ends of a magnet are called
the poles.
They are the best part to use
for attracting things.

This is how you can use a magnet
to keep your table neat.
Fix a magnet onto your table
with a lump of clay.
Which part of a pencil sharpener
do you think will be attracted?

What could you stick onto a pencil and
an eraser so that they are attracted
to the magnet?

# What things will a magnet work through?

Hide your magnet in a small box.
See if it will work through the box
to attract a paper clip.
Take your magnet out of the box.
Now wrap a piece of cloth
around your magnet.

Do you think it will still
attract a paper clip?
Find out if your magnet will work
through other things like this.

# Underwater Rescue

Make an underwater scene like this.
You need ★ A clear plastic container
        ★ 2 pieces of plastic cut from a detergent
          bottle (ask an adult to cut the two pieces
          of plastic for you)
        ★ some small stones or shells

Draw and cut out two divers
from the pieces of plastic.
Attach something to the back of each diver
so that they are attracted to the magnet.
Fill the container with water.
Make each diver sink to the bottom
of the sea.

Now they are trapped
between the rocks.
Can you use your magnet on the outside
of the container to rescue them?

What things did your
magnet work through to
rescue the divers?

Tie a piece of string around
the middle of a bar magnet
and hang it up.
Take some of the things that
a magnet will attract and
make long chains.
What do you think is holding
all these things together?

14

Put a nail file or a bobby pin
near a pin.
How could you make the nail file
pull the pin along?

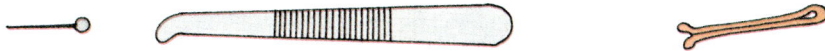

What happens to the nail file
when a magnet is near?

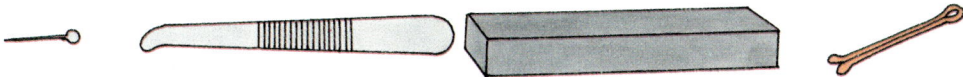

What happens to the nail file
when you take the magnet away?
Keep testing this.

Can you make the nail file
pull along any more pins?

# How to make your own magnet

Take the nail file in one hand and
a bar magnet in the other hand.
Stroke the nail file with the magnet.
You must always stroke the same way.
(It's like stroking a cat.)
Count 20 strokes.
See what your new magnet
will pick up.
Will it pick up more pins
than before?

Give your new magnet
another 20 strokes.
Does this make your magnet
pick up more things?

What happens if you bang
your magnet by mistake?

Test this on the magnet
you have just made.
Use a wooden block to bang
the nail file.
What happens when you try to
use it to attract things?

Be careful not to drop or bang
a real magnet.
You will make it weak and
it will not work so well.

# What happens when you put two magnets together?

Can you feel them pulling together?
They are attracting each other.
Can you feel them pushing away?
We say they are repelling
each other.

What happens when you put your
magnets together like this?

# The Fish and Chip Puzzle

You need ★ 2 bar magnets
        ★ 4 small pieces of paper
        ★ a pencil
        ★ some tape

Draw some fried fish on
two pieces of paper.
Draw some chips (french fries) on
the other two pieces.
Put your magnets together so that
the poles attract each other.
Stick 1 fish picture on 1 pole and
1 chip picture on the other pole.
Work out where to put the
other two pictures so that
fish and chips always go together.

# How to make Magnetic Boats

You need ★ a bowl of water
         ★ a packet of needles
         ★ some pieces cut
           from an egg carton
           or a ceiling tile

What to do:

1. You need to make each needle
into a magnet.
Keep the needles in
the packet and stroke them
with your magnet.
2. Take the needles out of
the packet.
Test that you have made
them magnetic.
What happens when you put two
needles together like this?
3. Push each needle through
a piece of ceiling tile.
Let your boats float
on the water.
Can you line them up
so that they attract each other?

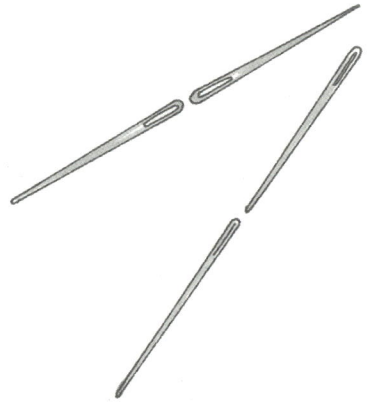

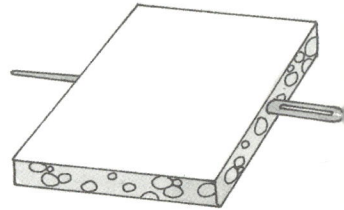

Take the needles out of
the boats.
This time push the needles
through the boats from top
to bottom.
The point of each needle
should be in the water.
Hold one pole of your magnet
over the boats.
What happens?

What do you think will happen
when you hold the other pole of
your magnet over the boats?

# The Swinging Magnet

Use paper and some thread
to make this sling for
a bar magnet.
Hang the magnet up
so that it can swing around.
Keep it away from anything made
of iron, like radiators or
metal table legs.
Do you know why?

Watch as the magnet stops
swinging.
Which way is it pointing?

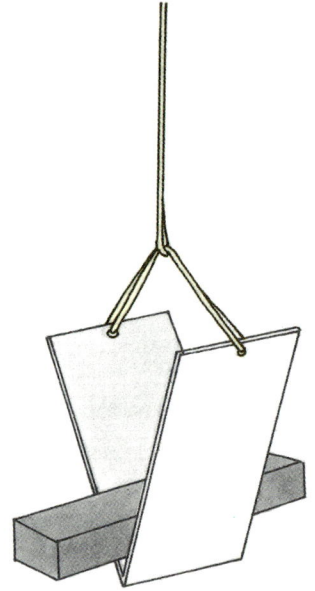

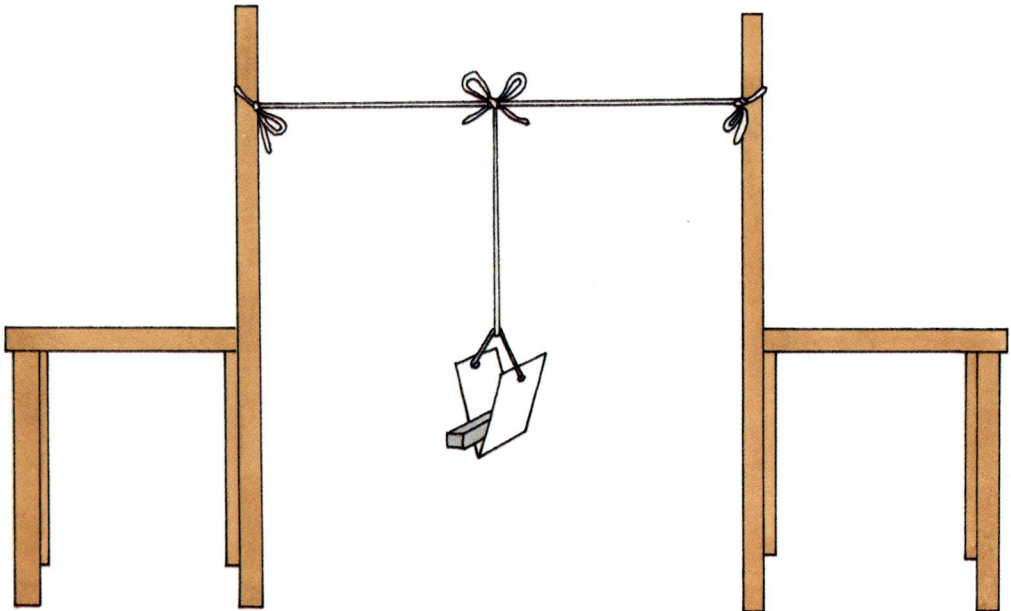

Make the magnet swing around again.
Watch as it stops.
Which way is it pointing
this time?
Does it always point in the
same way, or direction?

# The Floating Magnet

Put a bar magnet
into a plastic tub.
Make the tub float in a
bowl of water.
Watch as the tub stops
moving around.
Which way is the magnet
pointing?

Make the tub move around again.
Watch as it stops.
Does the magnet always point
in the same direction?

If you leave a magnet to
swing or float it will always
point in the same direction.

Now get a compass.
What do you notice about the direction
of your floating magnet and
the direction of the compass needle?

25

They are pointing
in the same direction.
The needle of a compass
is a magnet.
One pole of each magnet
is pointing to the North.
The other pole is pointing
to the South.
You use a compass to show you
which way you are going.
You have to turn the compass
so that the word North, or
the letter N, is underneath
the pole which points North.

# How to make your own Compass Card

You need ★ a piece of paper
      ★ a circle to draw around
      ★ a pencil, ruler, and scissors

What to do:

1. Trace around the circle on a piece of paper. Cut it out.
2. Fold the circle into half, then into quarters.
Use a ruler and pencil to draw along the fold lines.
3. Mark in North, South, East and West.
4. Check with the compass to see which pole points North. Put the compass under the Swinging Magnet or the Floating Magnet.
Make sure the North on the card is under the pole which points North.

Now you have made your own compass.
You could use it with a map to find out which direction you go to get home from school.

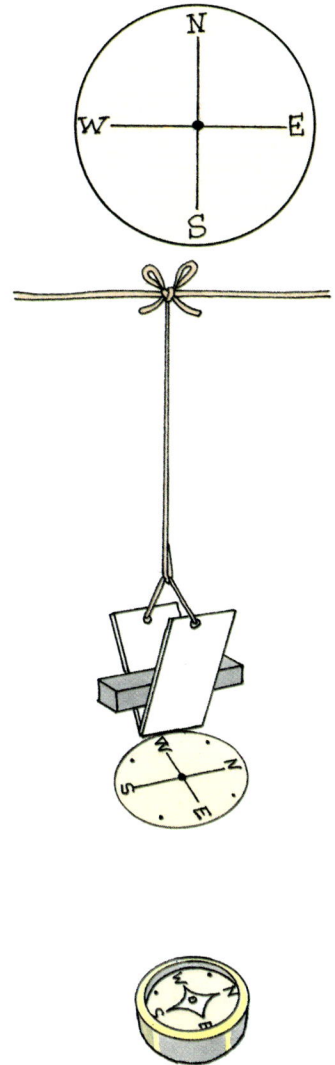

You may be using magnets
in your home.
You can use magnets to keep
notes or keys in place.
Magnets are used in refrigerators
to keep the door closed.

See if you can find some other places
where magnets are used.

music center

television

vacuum cleaner

Some magnets are made by
electricity.
They are called electro-magnets.
They are useful because
they can be switched on and off.
Electro-magnets are used to make
the motors work in these machines.

# Index

# Notes to parents and teachers

### Care and storage of magnets

Teach children to handle magnets properly. Tapping, hitting or dropping a magnet will weaken it. (The magnetic particles become misaligned.)

Make sure children know how to store magnets. They should be packed away, attracting each other, side by side. A keeper needs to be placed across the poles at each end of the magnets. This ensures that the lines of force form a continuous chain, keeping in the magnetism.

2 3 4 5 6 7 8 9 10—JDL—93 92 91 90 89 88 87